はしがき

　本書(ISBN978-4-938480-89-9)は、経済産業省、平成27年、発電用風力の規定に対応した技術改良の出版となった。その内容は規定に沿った技術からバードストライク（風車に鳥が衝突）を解決したもので、猛禽類、天然記念物でもあるオジロワシ、オオワシ、海ワシ、ハゲワシ、イヌワシ、トビなどが風車への衝突防止によるものである。猛禽類に対して、2段構えによるシステムであり、これをカラーイラストなどで、バードストライクのパターンについて解説し、画像、動画では、分かりにくい衝突、または激突する方向の寸前をイラストで解説した。

Preface

This book (ISBN 978-4-938480-89-9) will be publication of the technical improvement after regulation of the wind force for power generation in 2015 of the Ministry of Economy, Trade and Industry

The contents are the contents by the collision preventive measure of the bird by bird strike in the text.

To the bird, the contents were the systems by the two-step style, are color illustrations etc. and explained this about the pattern of bird strike.

目　次

(1) 経済産業省、平成27年発電用風力の規定後に対応した技術改良について‥6
　① 風力発電の高所から鳥の通過のパターン
　② 風力発電の中段から鳥の通過のパターン
　③ 高速で回転するブレードの先端部の動きが見えないモーションメスラ～
　④ 高速で回転するブレードの先端部の動きが見えるようにし、視認性～
　⑤ ブレードの先端部に取り付け設けられた素材は～

(2) 風車の前方の上段高所から進入する鳥‥‥‥‥‥‥‥‥‥‥‥‥‥‥‥‥‥7
　① 鳥が風車の最も高いところ１９０メートル前後のブレード間を通過する
　　 手前（イラスト文字解説）
　② 鳥がブレードの回転範囲内１－３を通過中（図２）
　③ ブレードの回転範囲内を通過中、鳥の背側方向からブレード１－２が接近
　④ 長さ９０ｍ前後のブレード１－２の先端が、鳥の背側に激突

(3) 風車の前方の中段から進入する鳥／中段からのバードストライク‥‥‥‥11
　① 鳥が中段のブレードに接近
　② 鳥がブレードとブレードの先端部の間に接近
　③ 鳥がブレードとブレードの先端部の間に接近
　④ ブレードが回転する範囲内の手前を通過中
　⑤ ⑥ ⑦　ブレードが回転する範囲内を通過中　ブレードが、鳥の背側に激突
　⑧ ⑨　ブレード２－１が、鳥の背側に激突、鳥が損傷を受け落下

(4) モーションメスラにより視認性不可‥‥‥‥‥‥‥‥‥‥‥‥‥‥‥‥‥20

(5) モーションメスラ回避による視認性‥‥‥‥‥‥‥‥‥‥‥‥‥‥‥‥‥21
　① 回転するブレードの先端部の視認性
　② 上段高所からの回転するブレードの先端部の視認性
　③ 風車の前方の中段から進入する鳥への回転するブレードの先端部の視認性
　④ ブレードの先端部に取り付け設けられた素材

(6) モーション・メスアの現象について‥‥‥‥‥‥‥‥‥‥‥‥‥‥‥‥‥25

(7) モーション・メスアの解消について

(8) ブレードにセットする素材について

(9) 各種ブレードの長さからコンピューターシュミレーションによる素材のサイズ
　　 ブレードの長さ／素材のサイズ　システム１　システム２

２、英語解説

Table of contents

(1) ···26

This has the thunderbolt by fixing an instrument to a windmill

(2) ···27

The bird which advances from the upper row height ahead of a windmill

　①

A bird passes a height of 190m of a windmill.

　②

A bird is passing [be / it] windmill's rotation within the limits 1-3.

　③

While passing windmill's rotation within the limits, a windmill 1-2 approaches [of a bird] from a back side.

　④

The tip of the windmill 1-2 before and behind a length of 90m crashes into the back side of a bird.

(3) ···31

The bird strike from [from the middle ahead of a windmill] the bird which advances / middle

　①

A bird approaches the windmill of the middle.

　②

A bird approaches between the tip parts of a windmill and a windmill.

　③

A bird approaches between the tip parts of a windmill and a windmill

　④

It passes through this side within the limits which a windmill rotates.

　⑤

It passes through this side within the limits which a windmill rotates.

　⑥

It passes through this side within the limits which a windmill rotates.

⑦

Passage and a windmill 2-1 crash into the back side of a bird within the limits which a windmill rotates.

⑧

A windmill 2-1 crashes into the back side of a bird.

⑨

A windmill 2-1 crashes the back side of a bird, and a bird receives damage and falls.

(4) --40

Visibility by Motion Mesa

(5) --41

Visibility by Motion Mesa evasion

①

Visibility of the tip part of the rotating windmill

②

Visibility of the tip part of the windmill which rotates from an upper row height

③

Visibility of the tip part of the rotating windmill from the middle ahead of a windmill to the bird which advances

④

The material attached and prepared in the tip part of a windmill

(6) --45

About the phenomenon of Motion Mesa

(7)

About the dissolution of Motion Mesa

(8)

About the material set to a windmill

(9)

Size of the material by the computer simulation from the length of various windmills, the length of a windmill

(1) 経済産業省、平成27年、発電用風力の規定後に対応した技術改良について

　　平成27年2,6の規定では、雷対策が追加された。これはバードストライク（鳥が風車に衝突）を避ける方法として、風車ブレードなどにステンレス製のベルトなどで器具を固定することによる落雷などもあり、また、伝導性の材質などでもあり、この技術では規定に沿った保安水準による確保の達成とは考えられない。本書では雷対策を踏まえた風車への雷撃の電荷量を３００クーロン～６００クーロン以上を想定したレセプターを設けた技術に支障を与えないシステムで規定に適合するものである。

① 風力発電の高所から鳥の通過のパターン

　　高さ150m前後、発電量３～８MW位の規模における風力発電のバードストライクを(2)、7～10頁、図1～図4においてイラストで解説した。

② 風力発電の中段から鳥の通過のパターン

　　高さ150m前後、発電量３～８MW位の規模における風力発電のバードストライクを(3)、11～19頁、図5～図13においてイラストで解説した。

③ 高速で回転するブレードの先端部の動きが見えないモーションメスラは、(4)、20頁、図14においてイラストで解説した。

④ 高速で回転するブレードの先端部の動きが見えるようにし、視認性を高めたブレード、モーションメスラの回避は、(5)、21頁、図15においてイラストで解説した。

⑤ ブレードの先端部に取り付け設けられた素材は、省令、風車、第4条,2、風圧に対して構造上安全であること。5、運転中に他の工作物、植物等に接触しないように施設すること。等に適合したシステム１～２である。

※宣伝・説明・明細書表現などへの引用を禁止。

(2) 風車の前方の上段高所から進入する鳥／上段高所バードストライク
　① 鳥が風車の最も高いところ１９０メートル前後のブレード間を通過する手前９０メートル前後（図１）

図１

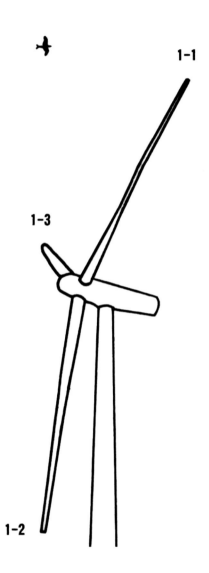

図１は、鳥が１－１と１－３のブレードの１６０ｍ前後の空間を通過しようとブレードに接近する。この場合、ゆっくり回転しているブレードの間を通過しようと飛行する。

※宣伝・説明・明細書表現などへの引用を禁止。

② 鳥がブレードの回転範囲内1-3を通過中(図2)

図2

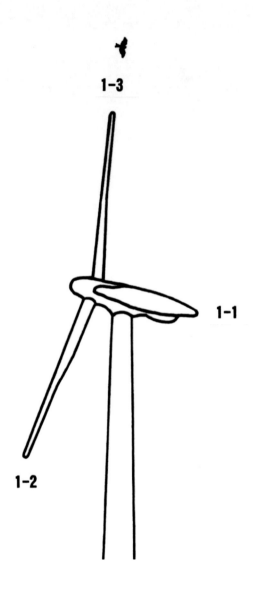

この場合、ブレードの先端部は、時速300キロ前後で回転しているため、ブレードの先端は見えない。これが「モーション・メスア」という現象でもある。これにより、ブレードが、鳥に激突するのは、背側からが多く見られる。(図2)

※宣伝・説明・明細書表現などへの引用を禁止。

③ ブレードの回転範囲内を通過中、鳥の背側方向からブレード１－２が接近

図３

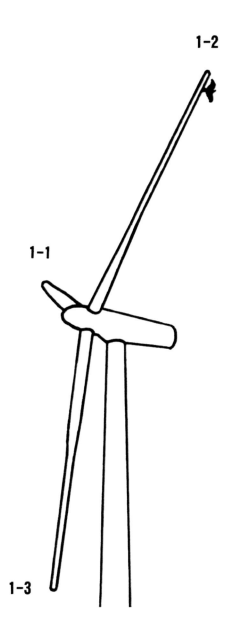

図３は、鳥の背側方向から高速で接近するブレード１－２が背側に激突する寸前であるが、上から下に回転するブレードに衝突する鳥が多い。

※宣伝・説明・明細書表現などへの引用を禁止。

④ 長さ９０ｍ前後のブレード１－２の先端が、鳥の背側に激突

図４

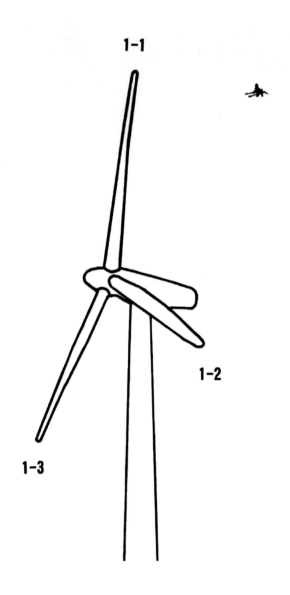

図４は、背側から激突されて鳥がダメージを受け、落下するものであるが、鳥の損傷は、殆どが骨が折れている。
これがモーション・メスアが原因と思われるバードストライクである。

※宣伝・説明・明細書表現などへの引用を禁止。

⑶ 風車の前方の中段から進入する鳥／中段からのバードストライク
　① 鳥が中段のブレードに接近

図５

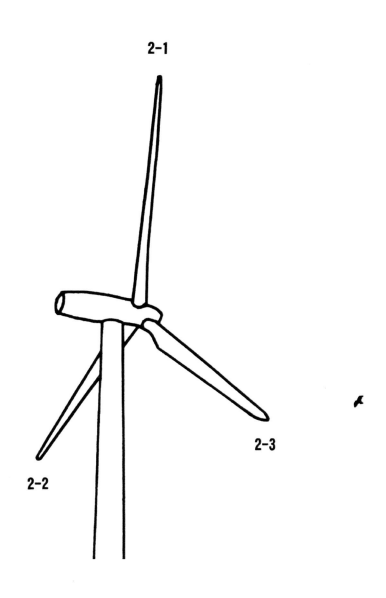

図５は、鳥がブレードとブレードの間を通過しようと風車の約１００メートル手前に接近。

※宣伝・説明・明細書表現などへの引用を禁止。

② 鳥がブレードとブレードの先端部の間に接近

図6

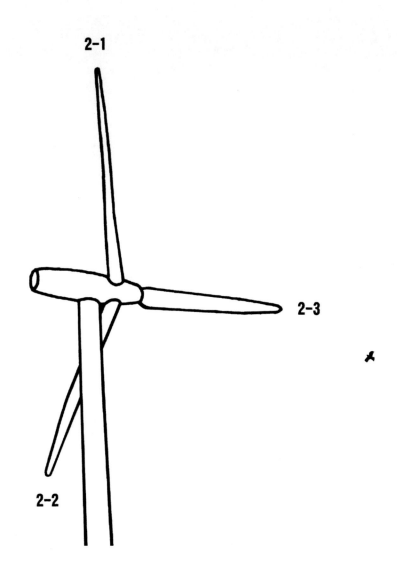

図6は、鳥がブレードとブレードの間を通過しようと風車の約60メートル手前に接近。

※宣伝・説明・明細書表現などへの引用を禁止。

③ 鳥がブレードとブレードの先端部の間に接近

図7

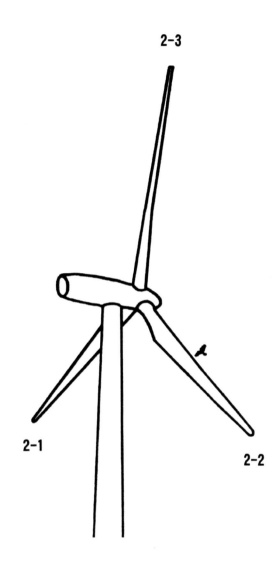

図7に示す。鳥がブレードとブレードの間を通過しようと風車の約30メートル手前に接近。

※宣伝・説明・明細書表現などへの引用を禁止。

④ ブレードが回転する範囲内の手前を通過中

図8

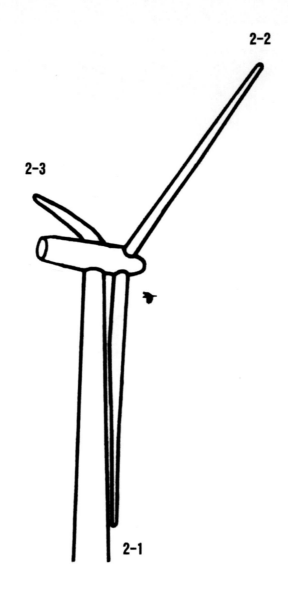

図8は、鳥がブレードとブレードの間を通過しようと風車の約10メートル手前に接近。

※宣伝・説明・明細書表現などへの引用を禁止。

⑤ ブレードが回転する範囲内を通過中

図9

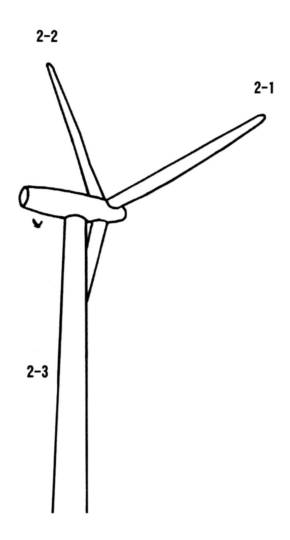

図9は、鳥がブレードとブレードの間を通過しようと風車の回転範囲内ブレードの約2メートル手前に接近。

※宣伝・説明・明細書表現などへの引用を禁止。

⑥ ブレードが回転する範囲内を通過中

図 10

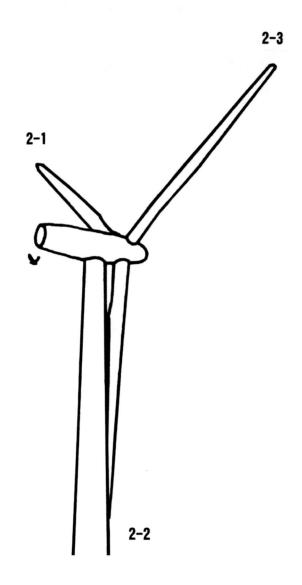

図 10 は、鳥がブレード 2－1 とブレード 2－2 の間を通過しようと風車の回転範囲内ブレードの約 1 メートル手前に接近。

※宣伝・説明・明細書表現などへの引用を禁止。

⑦ ブレードが回転する範囲内を通過中、ブレード２－１が、鳥の背側に激突
図11

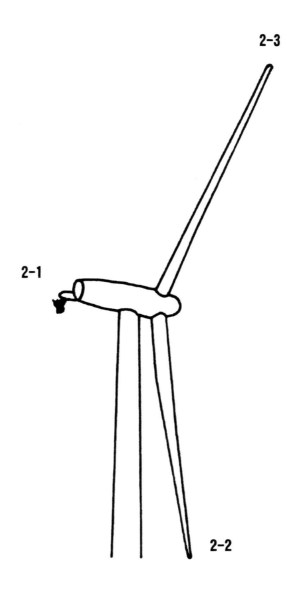

図11 は、ブレード２－１の先端方向の回転速度は、時速 300 キロ前後であるため、通過の手前では視認性（目による障害物の確認）の作用が困難であり、ワシの通過の速度は、時速 45-65 キロ位であり、回転範囲内の衝突を避けるのは難しい。これがモージョン・メスアが原因と思われるバードストライクである。

※宣伝・説明・明細書表現などへの引用を禁止。

⑧　ブレード2-1が、鳥の背側に激突

図 12

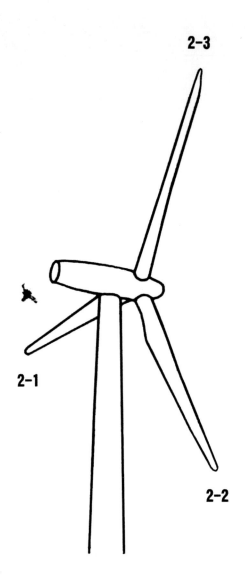

鳥は背側からブレードが激突されることが多い。この原因は背側から高速で接近するブレードの視認性が困難であることは明らかである。(図 12)

※宣伝・説明・明細書表現などへの引用を禁止。

⑨ ブレード2-1が、鳥の背側を激突し、鳥が損傷を受け落下

図13

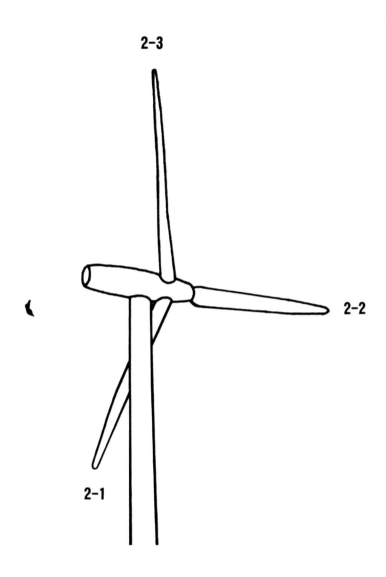

図13は、鳥の背側からブレード2-1が時速約300キロ前後の猛スピードで激突するから激突された鳥は、空中に浮いた状態になり、落下する。

※宣伝・説明・明細書表現などへの引用を禁止。

(4) モーションメスラにより視認性不可

図14

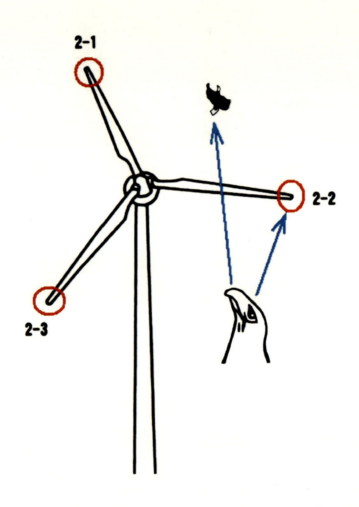

ブレードの先端部がモーションメスラで見えない為、ブレードが回転する範囲内を通過するのでブレードの衝突、バードストライクまたはブレードが背側から激突。前記の３図、１１図を参照。創作；IDF

※宣伝・説明・明細書表現などへの引用を禁止。

⑸ モーションメスラ回避による視認性
 ① 回転するブレードの先端部の視認性

図15

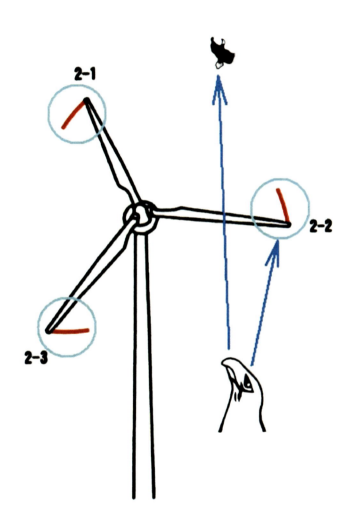

ブレードの先端部は見えているのでモーションメスラは回避され、ブレードの回転範囲を避けて通過。創作；IDF

※宣伝・説明・明細書表現などへの引用を禁止。

② 上段高所からの回転するブレードの先端部の視認性

図16

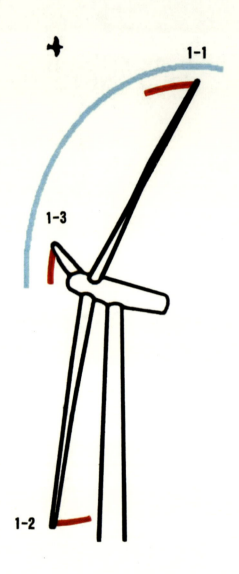

鳥が風車の最も高いところ190メートル前後のブレード間を通過する手前90メートル前後でブレードの先端部の視認性により、回転範囲を避けて通過。創作：IDF

※宣伝・説明・明細書表現などへの引用を禁止。

③ 風車の前方の中段から進入する鳥への回転するブレードの先端部の視認性

図17

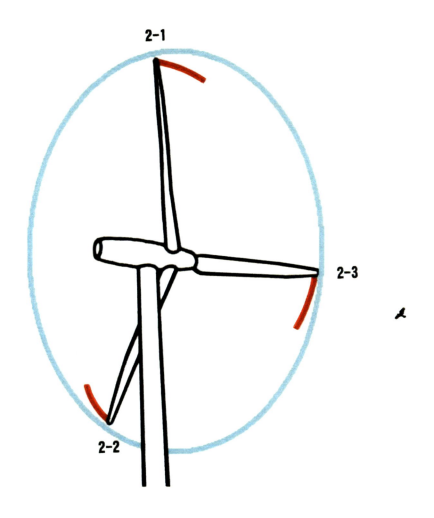

図17は、鳥がブレードとブレードの間を通過しようと風車の約100メートル手前で回転するブレードの先端部のリング状、或いは連続的に障害物の視認性により、ブレードの先端部を避けて通過。創作；IDF

※宣伝・説明・明細書表現などへの引用を禁止。

④　ブレードの先端部に取り付け設けられた素材

図18

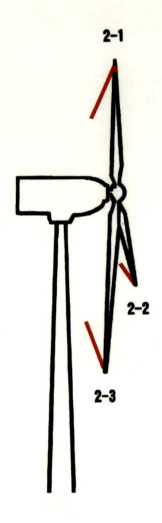

図18は、風速5mで発電するブレードの先端部2-1、2-2、2-3に斜めに表示されている素材は、取り付け設けられた素材であり垂直方向の角度は約24度であり、これは素材の長さと関係する。

図15, 16, 17に示すモーションメスラ回避のシステム1の次のシステム2は、鳥の進入を素材が回転力で物理的に阻止し、ブレードの回転範囲を避ける。創作；IDF

※宣伝・説明・明細書表現などへの引用を禁止。

⑹ モーション・メスアの現象について
風力発電の風車のブレードとブレードの間が広く空いている為、その間を鳥、猛禽類（タカ、ワシなど、カラスを除く）が、通り抜けようとし、ブレードに衝突するものである。　ブレードの先端方向の回転速度は、時速300キロ前後であるため、通過の手前では視認性（目による障害物の確認）の作用が困難であり、ワシの通過の速度は、時速45-65キロ位であり、風車の回転速度による衝突を避けるのは難しい場合が多い。　しかし、猛禽類のハヤブサは時速200キロ以上であるなら通過できるでしょう。　このモージョン・メスアが原因と思われるバードストライク、鳥類の衝突死で目立つのが、天然記念物でもあるオジロワシ、トビなどであり、オオゼロカモメ、カラス類の順位となっている。

⑺ モーション・メスアの解消について
ブレードの先端方向の視認性を高める構成は、連続的に見えるものでもあり、また、連鎖性を有するものでもある。このリング状、連続的、連鎖性の作用が視認性に効果を発揮するシステム1である。　図15,16,17に示すように、ブレードの先端部にロープ状、或いは、帯状、テープ状から成る素材を取り付けて設ける。
上記のシステム1の他に別な効果によるシステム2もある。その効果・作用は、ブレードとブレードの間を通過する際に通過が阻止される構成となる。その構成はロープ状、帯状、テープ状の側面が進入を阻止するものであり、また忌避剤を塗布したものである。これにより鳥の背側へのブレードの衝突、バードストライクを避けるものとなる。　図18

⑻ ブレードにセットする素材について
ブレードの長さ、25m，45m，63m，93m，100mなどに適した素材、形状、構成があり、これはPCのシュミレーションなどによるものである。

⑼ 各種ブレードの長さからコンピューターシュミレーションによる素材のサイズ

　　　ブレードの長さ／素材のサイズ　システム1　システム2

　1，　25m前後／　　5～10m前後　　2～10m前後

　2，　45m前後／　　9～14m前後　　2～14m前後

　3，　63m前後／　12～22m前後　　2～22m前後

　4，　93m前後／　17～30m前後　　2～30m前後

　5，100m前後／　20～40m前後　　2～40m前後

ブレードの幅、42cm前後～62cm前後　レセプターを含む場合もあり（上記1，2）、素材のセッティングは、素材とブレードの長さにより異なり、風力を考慮した形状になる。従って、これがブレードの長さに対応した素材サイズのマニュアル。作成：IDF

※宣伝・説明・明細書表現などへの引用を禁止。

(1)
This has the thunderbolt by fixing an instrument to a windmill etc. by the belt made from stainless steel etc. in a bird strike dissolution, and is also the conductive quality of the material etc., and is not considered to be achievement of the reservation by the security level in alignment with regulation with this technology.
It is the system which does not give trouble to the technology which prepared the receptor supposing the amount of electric charges of the thunderstroke to the windmill based on the measure against thunder in this book.

①
The pattern of passage of the height of wind power to a bird
In 7-10 page [of bird strikes of the wind power in the scale of the amount grade of 3-8 MW of power generation], Fig. 1 - figure 4, the illustration explained before and after a height of 150m.

②
The pattern of passage of the middle of wind power to a bird
In 11-19 page [of bird strikes of the wind power in the scale of the amount grade of 3-8 MW of power generation], Fig. 5 - figure 13, the illustration explained before and after a height of 150m.

③
In 20 pages and Fig. 14, the illustration explained Motion Mesa whose motion of the tip part of the windmill which rotates at high speed cannot be seen.

④
It is made seen [a motion of the tip part of the windmill which rotates at high speed], and the illustration explained evasion of the windmill which raised visibility, and Motion Mesa in 21 pages and Fig. 15.

⑤
The material attached and prepared in the tip part of a windmill should be structure top safety to a ministerial ordinance, a windmill, the 4th article, 2, and wind pressure.
They are the systems 1-2 which suited instituting so that other structures, a plant, etc. may not be contacted during 5 and operation.

The quotation to advertisement, explanation, specification expression, etc. is forbidden.

(2)
The bird / upper row height bird strike which advances from the upper row height ahead of a windmill

Before or after 90m of this side where a bird passes [highest] through between the windmills around 190m despite a windmill

Fig. 1

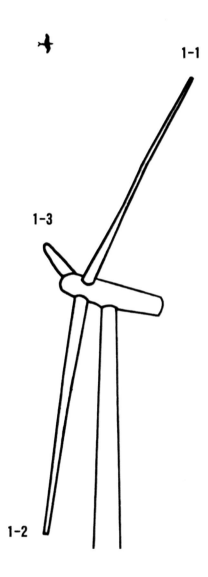

Fig. 1 approaches a windmill in order that a bird may pass through the space around 160m of the windmill of 1-1 and 1-3.
In this case, it flies to pass through between the windmills which are rotating slowly.

The quotation to advertisement, explanation, specification expression, etc. is forbidden.

②
A bird is passing [be / it] windmill's rotation within the limits 1-3.

Fig. 2

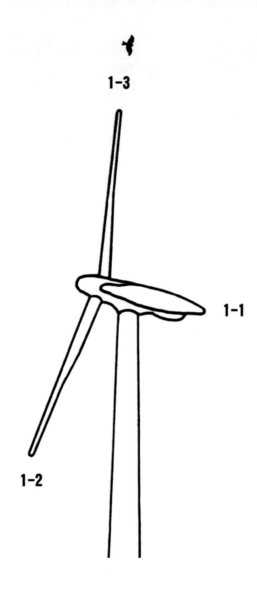

In this case, since the tip part of a windmill is rotating before and after 300km/h, the tip of a windmill is not visible.
This is also a phenomenon of "Motion Mesa."
Thereby, a windmill is seen crashing into a bird in many cases from a back side.

The quotation to advertisement, explanation, specification expression, etc. is forbidden.

③
While passing windmill's rotation within the limits, a windmill 1-2 approaches [of a bird] from a back side.

Fig. 3

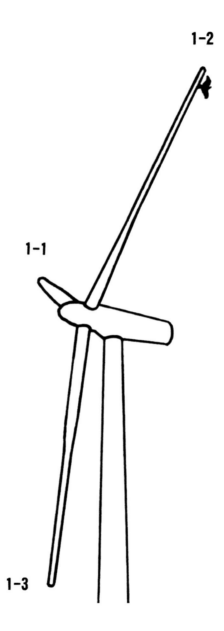

Although it is just before the windmill 1-2 which approaches [of a bird] at high speed from a back side crashes into a back side, there are many birds which collide with the windmill which rotates from a top to the bottom.

The quotation to advertisement, explanation, specification expression, etc. is forbidden.

④
The tip of the windmill 1-2 before and behind a length of 90m crashes into the back side of a bird.
Fig. 4

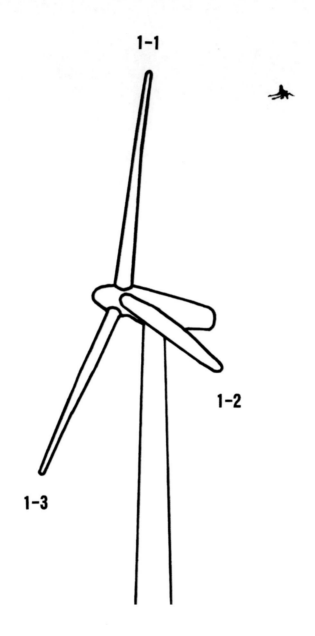

It crashed from the back side and the bird received the damage, and although it falls, as for the damage on a bird, the bone has broken [most].
This is the bird strike by which Motion Mesa is considered to be the cause.

The quotation to advertisement, explanation, specification expression, etc. is forbidden.

(3)
The bird strike from [from the middle ahead of a windmill] the bird which advances / middle
①
A bird approaches the windmill of the middle.

Fig. 5

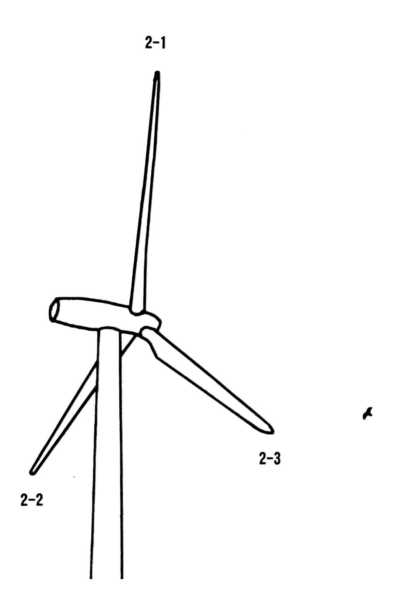

A bird approaches before [about 100m] a windmill to pass through between a windmill and windmills.

The quotation to advertisement, explanation, specification expression, etc. is forbidden.

②
A bird approaches between the tip parts of a windmill and a windmill.

Fig. 6

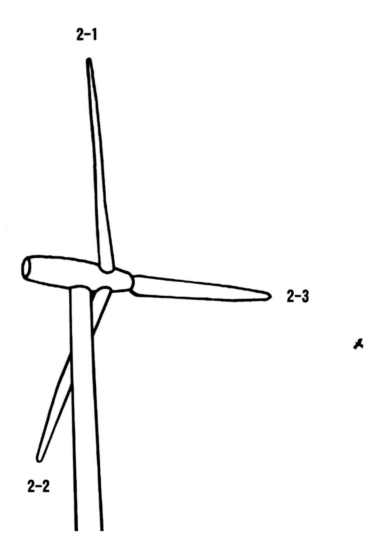

A bird approaches before [about 60m] a windmill to pass through between a windmill and windmills.

The quotation to advertisement, explanation, specification expression, etc. is forbidden.

A bird approaches between the tip parts of a windmill and a windmill.

Fig. 7

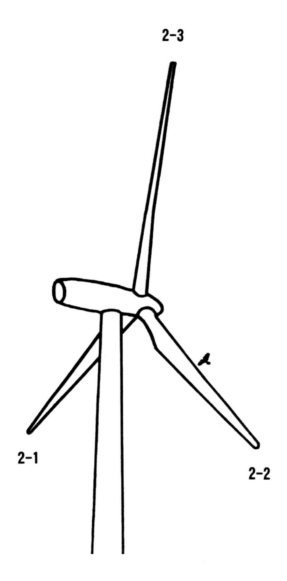

A bird approaches before [about 30m] a windmill to pass through between a windmill and windmills

The quotation to advertisement, explanation, specification expression, etc. is forbidden.

④
this side within the limits which a windmill rotates is under passage.

Fig. 8

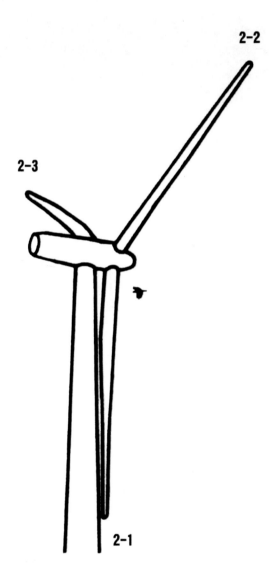

A bird approaches before [about 10m] a windmill to pass through between a windmill and windmills.

The quotation to advertisement, explanation, specification expression, etc. is forbidden.

⑤
A bird is passing [be / it] within the limits which a windmill rotates.

Fig. 9

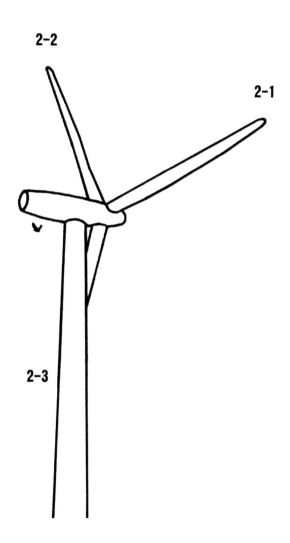

A bird approaches before [about 2m] the rotation within the limits windmill of a windmill to pass through between a windmill and windmills.

The quotation to advertisement, explanation, specification expression, etc. is forbidden.

⑥
A bird is passing [be / it] within the limits which a windmill rotates.

Fig. 10

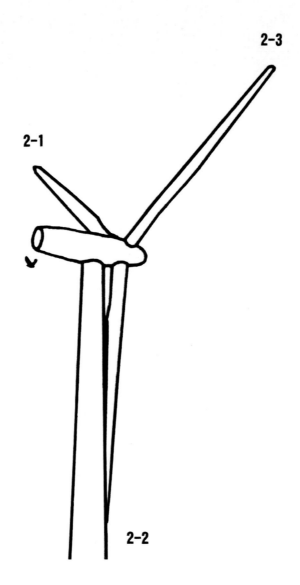

A bird approaches before [about 1m] the rotation within the limits windmill of a windmill to pass through between a windmill 2-1 and windmills 2-2.

The quotation to advertisement, explanation, specification expression, etc. is forbidden.

While passing within the limits which a windmill rotates, a windmill 2-1 crashes into the back side of a bird.

Fig. 11

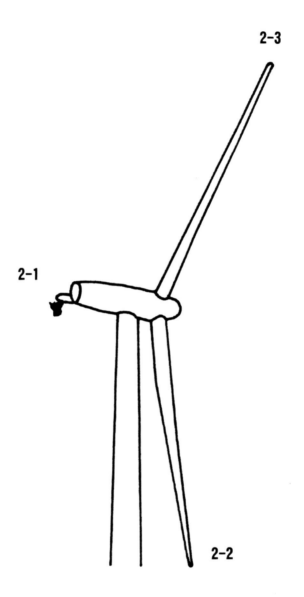

Since the rotation speed of the direction of a tip of a windmill 2-1 is before and after 300km/h, before passage, the action of visibility (check of the obstacle by the eye) is difficult, and the speed of passage of an eagle is speed per hour 45 - 65km grade, and is difficult for avoiding the collision of rotation within the limits.
This is the bird strike by which Motion Mesa is considered to be the cause.

The quotation to advertisement, explanation, specification expression, etc. is forbidden.

⑧
A windmill 2-1 crashes into the back side of a bird.

Fig. 12

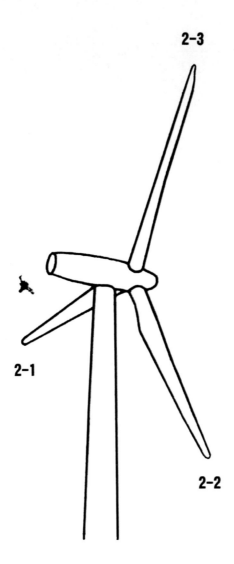

It crashes into a windmill to a bird in many cases from a back side.
This cause that the visibility of the windmill which approaches at high speed
is difficult is clear from a back side.

The quotation to advertisement, explanation, specification expression, etc.
is forbidden.

⑨

A windmill 2-1 crashes the back side of a bird, and a bird receives damage and falls.

Fig. 13

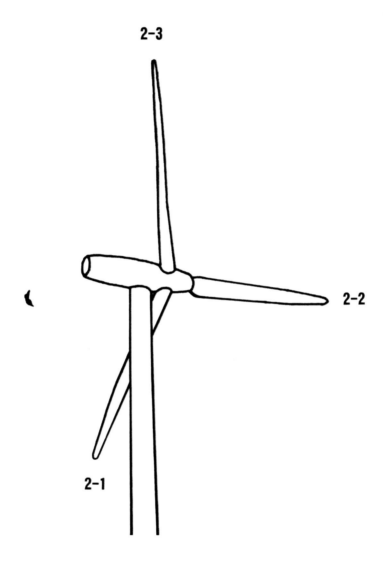

The bird into which it crashed since the windmill 2-1 crashed at the 猛 speed before and behind the speed of about 300km from the back side of a bird will be floated in the air, and will fall

The quotation to advertisement, explanation, specification expression, etc. is forbidden.

(4)
Visibility by Motion Mesa

Fig. 14

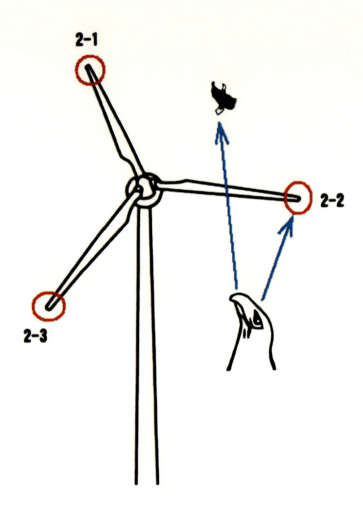

Since the tip part of a windmill cannot be seen by Motion Mesa and within
the limits which a windmill rotates is passed, the collision of a windmill,
a bird strike, or a windmill crashes from a back side.
The above-mentioned 3 figures and 11 figures are referred to.
Creation; IDF

The quotation to advertisement, explanation, specification expression, etc.
is forbidden.

(5)
Visibility by Motion Mesa evasion
　①
Visibility of the tip part of the rotating windmill

Fig. 15

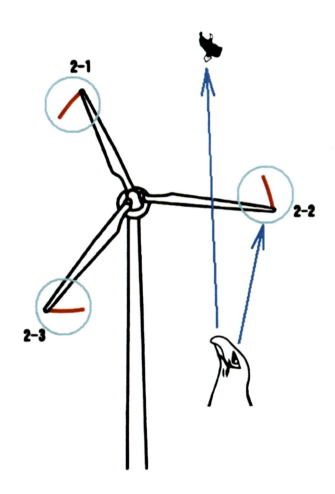

Since the tip part of a windmill is visible, Motion Mesa is avoided, avoids
the rotation range of a windmill and passes.
Creation; IDF

The quotation to advertisement, explanation, specification expression, etc.
is forbidden.

②
Visibility of the tip part of the windmill which rotates from an upper row height

Fig. 16

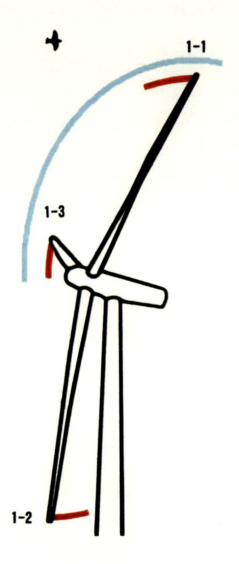

By the visibility of the tip part of a windmill, a bird avoids the rotation range before and behind 90m of this side which passes [highest] through between the windmills around 190m despite a windmill.
Creation; IDF

The quotation to advertisement, explanation, specification expression, etc. is forbidden.

Visibility of the tip part of the rotating windmill from the middle ahead of a windmill to the bird which advances

Fig. 17

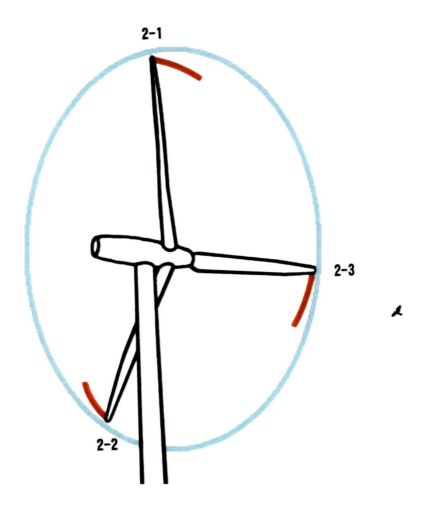

the shape of a ring of the tip part of the windmill which will rotate before
[about 100m] a windmill if a bird will pass through between a windmill and
windmills -- or -- continuous -- the visibility of an obstacle -- the tip
part of a windmill -- avoiding -- passage
Creation; IDF

The quotation to advertisement, explanation, specification expression, etc. is forbidden.

④
The material attached and prepared in the tip part of a windmill

Fig. 18

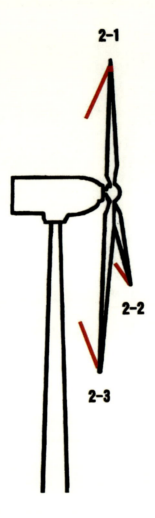

It is the material which the material currently displayed aslant attached in the tip part 2-1 of the windmill generated at 5m of wind velocity, 2-2, and 2-3, and was prepared in them, a perpendicular angle is about 24 degrees, and this is related to the length of a material.
A material prevents penetration of a bird physically on torque, and the next system 2 of the system 1 of Motion Mesa evasion shown 15, 16, and 17 avoids the rotation range of a windmill.
Creation; IDF

The quotation to advertisement, explanation, specification expression, etc. is forbidden.

(6)
About the phenomenon of Motion Mesa
Since between the windmill of wind power and windmills is widely vacant, a bird makes the meantime the method of passing through, and it collides with a windmill.
Since the rotation speed of the direction of a tip of a windmill is before and after 300km/h, before passage, the action of visibility (check of the obstacle by the eye) is difficult, and the speed of passage of an eagle is speed per hour 45‑65km grade, and is difficult for avoiding the collision by the rotation speed of a windmill in many cases.
However, if Peregrine is 200km [or more]/h, it will be able to be passed.
Haliaeetus albiclla, a kite, etc. which are also a natural treasure are conspicuous by the bird strike and the collision death of birds by which this Motion Mesa is considered to be the cause, and it serves as ranking of a sea gull and crows.

(7)
About the dissolution of Motion Mesa
The composition which raises the visibility of the direction of a tip of a windmill also looks continuous, and also has chain nature.
The actions of chain nature are the shape of this ring, and continuous and the system 1 which demonstrates an effect to visibility.
As shown Fig. 15, 16, and 17, the material which consists of the shape of the shape of a rope, a belt, and a tape is attached and prepared in the tip part of a windmill.
There is a system 2 by another effect besides the above-mentioned system 1.
Its effect and action are the composition that passage is prevented, in case it passes through between a windmill and windmills, and as for the composition, the side of the shape of a rope, a belt, and a tape prevents penetration.
Moreover, evasion paint is carried out, a damage is given to the eye of a bird, and penetration is prevented.
This avoids the collision of the windmill by the side of the back of a bird, and a bird strike.
This calls off a collision and bird strike of a windmill to the back side of a bird.
Fig. 18

(8)
About the material set to a windmill
There are a material suitable for the length of a windmill, 25m, 45m, 63m, 93m, 100m, etc., form, and composition, and this is based on the simulation of PC etc.

(9)
Size of the material by the computer simulation from the length of various windmills
1, 25m 5〜10m System 1 2〜10m System 2 Following same as the above
2, 45m 9〜14m   3, 63m 12〜22m   4, 93m 17〜30m
5, 100m 20〜40m
Before or after width [of a windmill], and 42cm order・62cm   A receptor may be included, and the arrangement of a material changes with length of a material and a windmill, and becomes the form where wind force was taken into consideration.   Creation; IDF
The manual of the material size corresponding to the length of a windmill

あとがき

風力発電開発事業に携わる方々との交流があった事が発案のきっかけとなりました。
この度、本書が発刊されました事は誠に意義深く感謝に堪えない次第です。

株式会社　ＵＶＣシステム研究所
代表取締役　樋口節美
810-0001
福岡市中央区天神１丁目９番１７号
福岡天神フコク生命ビル１５階
ＴＥＬ０９２－７１７－３９０３
ＦＡＸ０９２－７１７－３９０９
本件事業のお問い合わせ先

本書の奥付の発行№.について

№.が記載された剥離紙が、枠内に貼られ角印が押されている。従って、これが複写された本書は、内容が異なる海賊版でありますからご注意ください。

About issue [of the colophon of this book] No.

The angle mark is stamped on the seal with which No. was indicated. Therefore, since this book with which this was copied is a pirate edition from which the contents differ, it should be careful.

風力発電 鳥の衝突防止 バードストライクのパターンを回避したシステム

定価（本体 8,500 円＋税）

２０１７年（平成２９年）２月１５日発行

No.

発行所　IDF（INVENTION DEVLOPMENT FEDERATION）
発明開発連合会®

メール 03-3498@idf-0751.com　www.idf-0751.com

電話 03-3498-0751㈹

150-8691 渋谷郵便局私書箱第２５８号

発行人　ましば寿一

著作権企画　IDF 発明開発(連)

Printed in Japan

著者　樋口 節美©
　　　　（ひぐちせつみ）

本書の一部または全部を無断で複写、複製、転載、データーファイル化することを禁じています。
It forbids a copy, a duplicate, reproduction, and forming a data file for some or all of this book without notice.